House of Hidayah

A Children's Learning Enterprise

Copyright © 2019 House of Hidayah

Printed in the United States of America

On February 21, 1990, NASA was getting ready
to launch a space shuttle into our sky.
The space ship was huge and powerful.
It also had a beautiful name.
It was called Atlantis!

BUT THE DAY BEFORE THE TAKEOFF, NASA NOTICED
THAT THEY HAD A PROBLEM!
    THE SPACESHIP WAS FINE BUT ONE OF THE ASTRONAUTS
WHO WAS SUPPOSED TO BE ON THE ROCKET GOT SICK.

The name of this Astronaut was John Creighton.

He was a great sailor, He was also an expert in
flying planes and building rockets.Astronaut John
flew airplanes for over 6,000 hours!
But today, he was feeling tired.

Astronaut John Creighton caught a common cold.
He had a sore throat.
And he also had a runny rose.
Oh, he was not feeling very well!

His friends tried to help him.
"We will take care of you," they said.

Even though it was only a cold and not something too serious, NASA knew they couldn't send an astronaut into space when he had a cold.

What would happen if he had to sneeze while wearing a full body spacesuit? Astronaut John would not be able to wipe his nose if he was travelling in outer space! So, NASA had to wait for him to get better.

After one week, NASA had to keep the rest of the team waiting the whole time but Astronaut John finally got better!

The other astronauts took care of John. They brough him fruits. They got him candies and chips.

The other astronauts did not want to leave their friend behind on earth.

Little by little,
Astronaut John
started to feel better.

He no longer felt dizzy.
He did not feel tired.

Astronaut John put on his space suit.

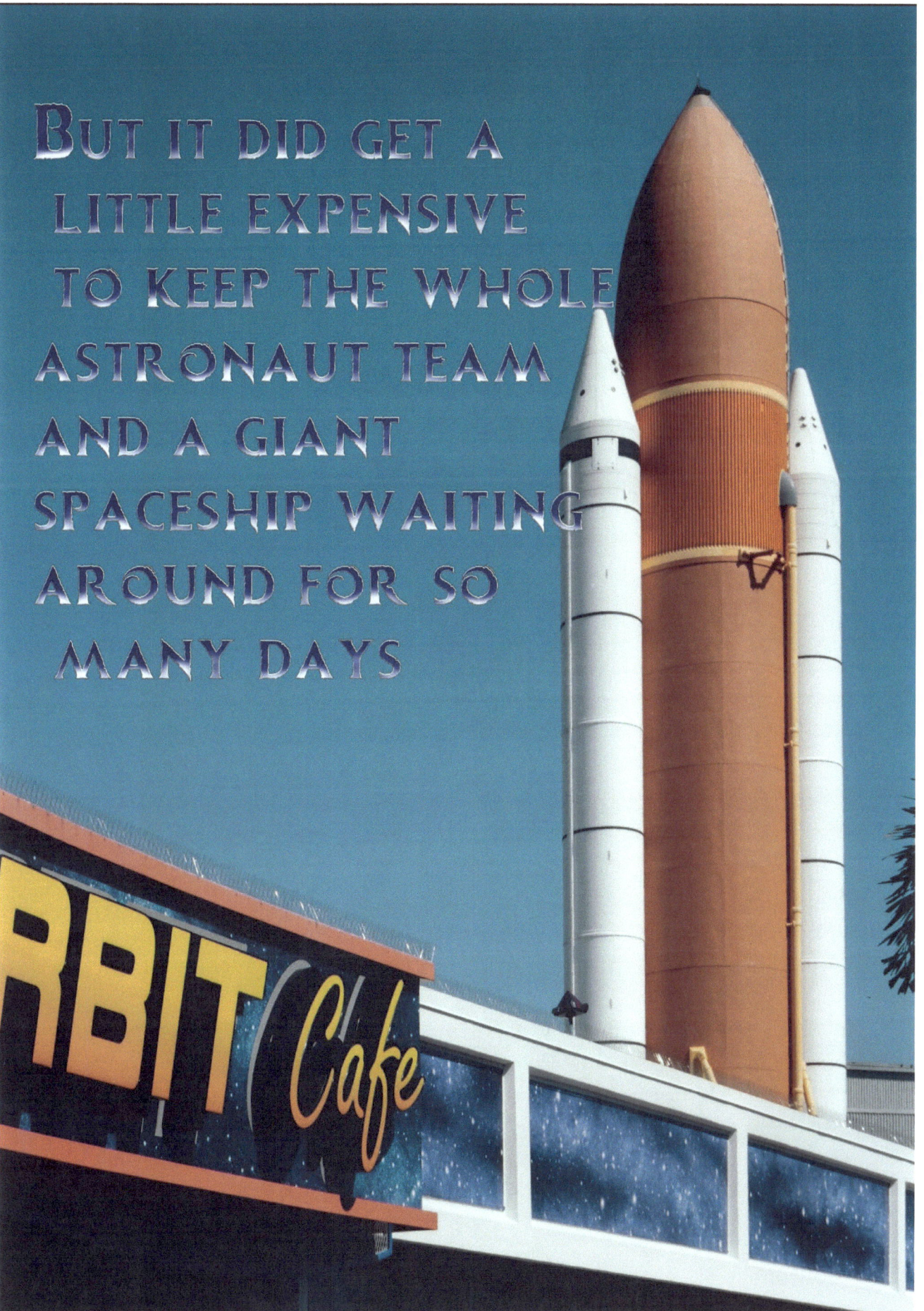

Astronauts John's cold was a
very expensive one!
It cost the NASA two and a half
million dollars!

BUT IN THE END ALL THE ASTRONAUTS WERE VERY HAPPY TO HAVE THEIR FRIEND BACK ON THE ROCKETSHIP ATLANTIS

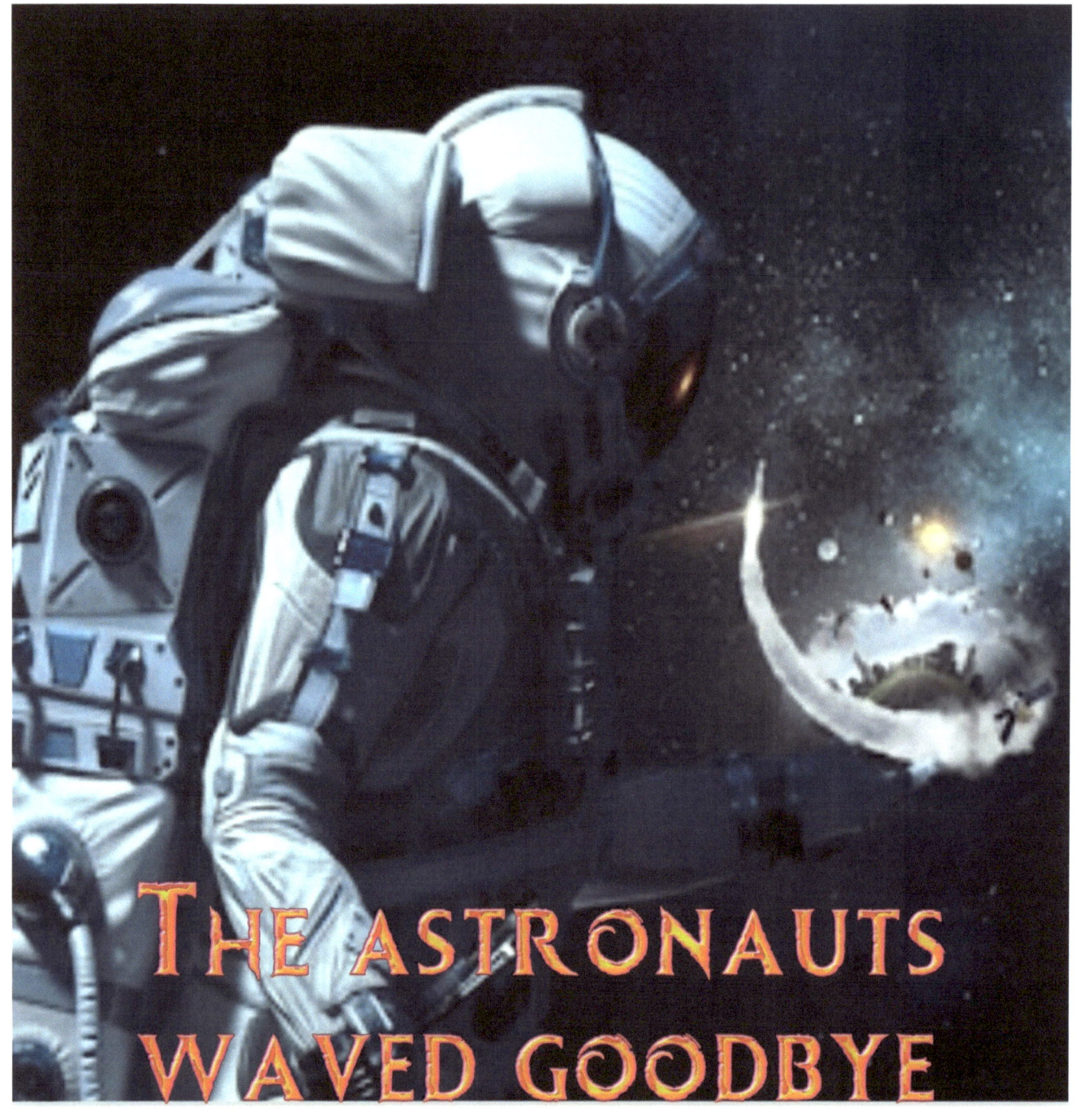

The astronauts waved goodbye to Earth

Then they took off to Space!!

Spaceship Atlantis roared away!

Name _____

Atlantis Spaceship

ACROSS

1   The name of a company which sends people to the moons and Space...
8   A sort of ship that takes people to The moon
9   A scientist who can go to the Moon...
10 When we feel very tired and weak
11 First name of the astronaut that got sick
12 Something that costs a lot
13 When the rocket pushes off the ground
15 When our body temperature goes up and we feel sick
16 The name of a spaceship that went to to Space

DOWN

2   Someone who studies about the Space
3   The planet we live on
4   A sort of plane that takes us to Space
5   A sort of ship that takes us to Space
6   A special suit you have to wear if you want to go to the Moon...
7   A place outside the world, where you can fins Stars, planets and moons...
14 Someone who you like and who likes you

Could you find these words
in the next page?

Atlantis
Spaceship
Space
Moon
Sun
Stars
Engineer
Pilot
Rocket
Fever
Million
Dollar
Space
Takeoff
Earth
NASA
Scientist
Real
Astronomy
Astronaut

Atlantis

Spaceship

```
N N Q S T R M N Y W R W D Y
A R S O C I O M L E D F W N
S G L P L I O C E A R N Q K
A I T L A N E N K I E D K T
P T I U O C I N E E O R L J
F O L R A G E N T L T N U S
N F T A N N D S L I R Q M R
E S O E N S O A H E S H B R
A C R E R T R R V I T T M R
G W A A K J I E T R P T D M
T W Z P T A F S A S M O O N
P J G D S S T E N L A N T Z
```

Atlantis

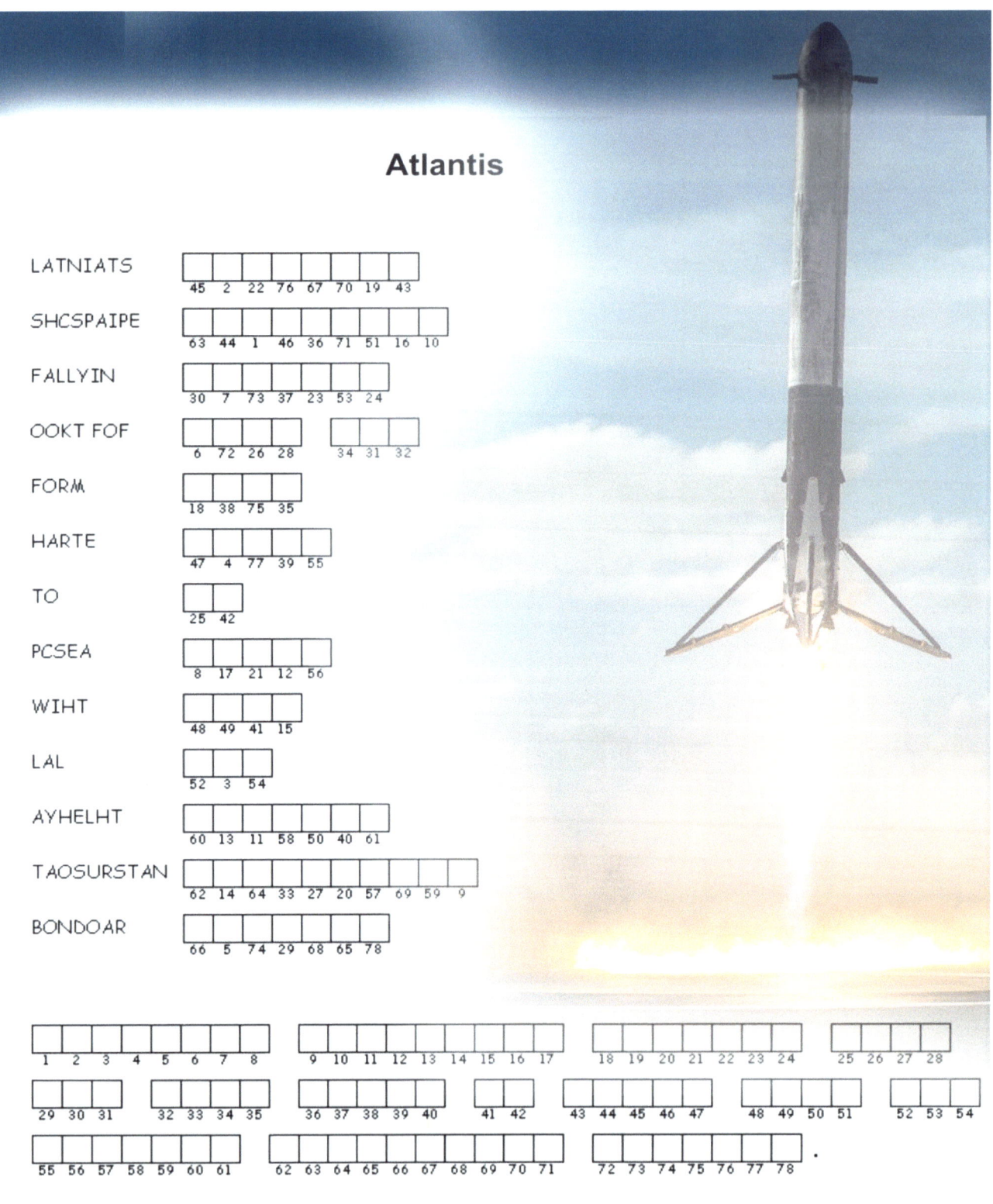

LATNIATS

45	2	22	76	67	70	19	43

SHCSPAIPE

63	44	1	46	36	71	51	16	10

FALLYIN

30	7	73	37	23	53	24

OOKT FOF

6	72	26	28	34	31	32	

FORM

18	38	75	35

HARTE

47	4	77	39	55

TO

25	42

PCSEA

8	17	21	12	56

WIHT

48	49	41	15

LAL

52	3	54

AYHELHT

60	13	11	58	50	40	61

TAOSURSTAN

62	14	64	33	27	20	57	69	59	9

BONDOAR

66	5	74	29	68	65	78

| 1 | 2 | 3 | 4 | 5 | 6 | 7 | 8 | | 9 | 10 | 11 | 12 | 13 | 14 | 15 | 16 | 17 | | 18 | 19 | 20 | 21 | 22 | 23 | 24 | | 25 | 26 | 27 | 28 |

| 29 | 30 | 31 | | 32 | 33 | 34 | 35 | | 36 | 37 | 38 | 39 | 40 | | 41 | 42 | | 43 | 44 | 45 | 46 | 47 | | 48 | 49 | 50 | 51 | | 52 | 53 | 54 |

| 55 | 56 | 57 | 58 | 59 | 60 | 61 | | 62 | 63 | 64 | 65 | 66 | 67 | 68 | 69 | 70 | 71 | | 72 | 73 | 74 | 75 | 76 | 77 | 78 | . |

Unscramble each of the clue words.
Copy the letters in the numbered cells to other cells with the same number.

Acknowledgments

This book would have been significantly incomplete without the magnificent illustrations of Mr. Adnan Oktar also known as Harun Yahya. (https://www.harunyahya.com/). We are grateful to his generosity in providing a variety of children-friendly images for use in this book.

Many special thanks to the illustrators from Unsplash including Hamid Khaeghi, Adam Miller, Sam Ti Kiefte, Spacex, Jason Mccann, Mike Strachwsky, Brian Minear and Nicolas Gras on *unsplash* for the glamourous images of rockets, space and planets.

To everyone at the *Shutterstock* website including Studiostoks, Vectorpocket, Trigubova Irina, Drawn to be Wild, Sergey Nivens, Alones and Vladi333 who enabled us to use their images in creating learning materials for children and turning their images into stories, we also thank you immensely for helping us create the most pleasant rocket story ever using those beaming illustrations.

Also written by this author:

Adam and his Rocketship

The Astronaut on Atlantis

The Girl who didn't listen to her Mom

The Little Girl and her Mean Friends

The Princess and her Lost Tiara

Zayd's Field Trip to Mars

www.ingramcontent.com/pod-product-compliance
Lightning Source LLC
Chambersburg PA
CBHW041311180526
45172CB00003B/1052